The Vegetable Files

A Lasting Record of Your Vegetable Gardening Knowledge and Experiences

Property of:

____

____

ISBN: 978-1-968971-25-0

And God said, “Let the earth sprout vegetation, plants yielding seed, and fruit trees bearing fruit in which is their seed, each according to its kind, on the earth.” And it was so.

Ironstone Tennessee

Disclaimer and Terms of Use:

The author and publisher have made every effort to ensure the completeness of this record book. However, they do not guarantee that all information is error-free or entirely comprehensive.

While diligent efforts have been made to verify the content, the author and publisher are not responsible for any errors, omissions, or differing interpretations of the material presented. Any unintended references that may appear to slight individuals, businesses, or organizations are purely coincidental.

Copies of this book are available for book signings, special events, and retail gift shop purchases. For direct ship information, of copies of 10 or more, please contact the author at ironstone_tennessee@yahoo.com.

The Vegetable Files
A Lasting Record of Your Vegetable Gardening Knowledge and Experiences

ISBN: 978-1-968971-25-0

Printed in the United States of America.

First printing, 2025

Ironstone Tennessee

Calhoun Tennessee, 37309

www.ironstone-tennessee.com

Notes

Table of Contents

How to Use The Vegetable Files

Firstly, I want to **thank you** so much for choosing to purchase The Vegetable Files. I welcome feedback and suggestions for ways to make it more useful for future editions.

There are over a dozen page styles, each grouped together so that you can organize your plant files in the way that works best for you. When you're finished, you'll have created a valuable reference as you go forward, and a priceless heirloom to pass on to future generations of gardeners.

Important Dates and Data:

Record your **USDA growing zone, average and observed frost dates, average annual rainfall, soil type and pH here.** This information allows you to choose plants that will be most likely to do well in your climate. Most seed packets, plant labels and catalog listings will tell you the plant's range of growing zones.

Knowing your **average first and last frost dates** will help you to schedule appropriate shipping dates when purchasing live plants and trees online. It will also help you to schedule when to start seeds indoors, when it is (probably) safe to transplant outdoors, and when you will need to have raised beds and garden plots prepared for Spring and Fall planting.

Keeping in mind your **annual average precipitation** (snow and rain) will help you to choose what plants to grow based on your desired amount of time spent in the garden (and how big of a water bill you can afford). If you are a lazy gardener like me, or just don't have time in for your schedule for frequent watering, then choosing plants that can thrive with only rainfall (most of the time) makes sense.

Your **soil type and pH** will determine what amendments you may need, what gardening methods will be most successful for you, and what plants will and won't do well for you in the garden. A bit of quick Internet research will tell you whether a plant likes your type of soil and whether your soil pH is in the right range for it to thrive. Knowing this can save you a lot of money and work.

My Gardening Challenges:

These are factors that you can't change easily (or at all), such as a short growing season, your work schedule, soil type, physical limitations, small budget or climate. While it's

easy to think of these challenges as a big negative, I would encourage you to instead think of them as boundaries that help you to keep your focus. Knowing from the start that you can't do All. The. Things., you are free to focus instead on the things you can do, and then make a plan to do those things well. Which brings us to:

My Gardening Goals:

Are you crazy for hot peppers? Do you want to grow your own spaghetti sauce? Are you trying to save money at the grocery store? Maybe you want to grow and store a year's worth of sweet potatoes. Maybe your goal is to enjoy fresh salads and salsa all summer long. Having goals and keeping them in focus will help you to achieve the vegetable garden of your dreams without going broke in the process.

Harvest Months:

The Harvest Month pages allow you to see what each month's harvest schedule looks like, week-by-week. For weeks with a heavy harvest schedule, you could divide the lists into two columns.

Sections:

There are 14 different page designs that you can use to document your research and experience with growing vegetables. Although this workbook is designed for vegetable gardens (it's in the title!) I realize that most gardeners grow at least a few small fruits and herbs. I've included one section for fruit(section 8), and one for herbs(section 11) in the workbook. For most vegetable gardeners, the 12 pages in each of those sections will be sufficient for your needs. If you're a serious herb grower or have a decent-sized orchard, you'll likely find that the Herb Files or the Orchard Files will meet your needs better.

Common Name:

Just as it suggests, this is the name by which most people recognize a plant. Examples would be broccoli, tomato, bush beans or sweet potatoes.

Latin Name:

This is the name that botanists or horticulturists use to refer to the plant. These names are important to regular gardeners too. Often the Latin name will tell you something about the plant or what other plants are related. For example: cabbage, broccoli, brussels

sprouts, kale, turnips and collards are all in the brassica family. They often share similar growing needs and have similar pest and disease problems.

Cultivar:

Many plants come in a variety of sizes, shapes, growth habits, flavor profiles and colors. These plants usually are given a cultivar name to differentiate one from another. For example: There are "slicing" tomatoes, "paste" tomatoes, "cherry" tomatoes and "grape" tomatoes. While Chadwick, Celebrity, San Marzano and Sweet 100 are all red tomatoes, they have very different characteristics that matter to both the grower and the chef.

Type:

This might refer to growth habit such as determinant (bush) and indeterminate (long, vining) tomatoes. In the case of strawberries and onions, the term "day neutral" refers to the light requirement (or lack thereof) for producing fruit or bulbs. Heirloom, open-pollinated or hybrid refers to the way in which growers select and produce seed.

Harvest Time:

This can refer to the part of the season when the plant produces food. Some plants are harvested in early spring, some produce for a period of weeks in late spring, and some are harvested in the fall. This is an important consideration when you are planning your garden, as you don't want to overwhelm yourself with more harvest at one time than you can handle.

Propagation:

Do you grow your tomatoes from seed or do you buy started bedding plants (or maybe you use rooted cuttings from last year's crop)? Do you start your onions from seed or from purchased onion sets? Rhubarb, horseradish and asparagus are often started from bare-root plants or by dividing existing plants.

Source:

It's good to record where your plants come from for several reasons. If you decide you want more, you know where to go. If you aren't happy with a plant's performance, you'll know to try a different source next time. If someone asks you where you got that mild jalapeno plant, you can, if you choose, share your source.

Light Preference:

Most garden vegetables prefer full sun, but there are a few that either prefer shade, or that can prolong the growing season by providing shade once the days start getting hot. Before you put your plants in the ground, it's good to give some thought about where they will get the best light to grow.

Soil Type:

Some plants are very picky about the soil type they prefer. If your favorite plant doesn't favor your soil, you will need to decide just how far you are willing to go to have that plant. Your soil can be improved, but it takes time and work. In some cases, you might be better off growing those plants in containers.

Water Requirements:

Cucumbers can be gluttons for water. Knowing this can help you decide where to plant them (near the garden hose or by the back door). It can also help you decide what to plant. If you live in a drought-prone area and you pay for water, you might choose to limit how many of the water-loving vegetables you grow, or you may need to get creative with your mulching and watering.

Companion Plants:

It makes sense to plant things with similar light, soil and water requirements near each other. Beyond that, some plants benefit others by improving flavor, repelling pests or protecting them from disease.

Soil Amendments:

Some plants require the application of fertilizers or pH adjustment from time to time. Record the plant's needs here so you can find the information quickly.

Common Diseases:

Certain cultivars of plants are more or less disease prone than others. Some tomatoes are more susceptible to blossom end rot or fruit cracking when rainfall is uneven. When you know your plants, you're better able to keep watch and head off a small problem before it becomes a large problem. You're also able to decide whether the plants that are most susceptible to disease are worth your time to grow.

Pest Pressure:

Most plants have some kind of insect or animal that loves to feast on them. If you know what to watch out for, you can make a plan to prevent, treat or protect your plants as you see fit. You can also decide whether not it's worthwhile to grow those plants that are more vulnerable to pests.

Notes

The notes area is for your observations, ideas, plans, recollections and thoughts you don't want to forget. Things like "Only plant 2 zucchinis each year" or "needs a lot of room to spread" or "DO NOT plant near the potatoes".

I wish you Bountiful Gardening!

Important Dates and Data

Growing Zone	Soil Type	Soil pH	Avg. Last Frost	Avg. 1st Frost	Avg. Rainfall

My Gardening Challenges:

My Gardening Goals:

Harvest Times

Month/Week	January				February				March			
Plant	**1**	**2**	**3**	**4**	**1**	**2**	**3**	**4**	**1**	**2**	**3**	**4**

Harvest Times

Month/Week	April				May				June			
Plant	1	2	3	4	1	2	3	4	1	2	3	4

Harvest Times

Month/Week	July				August				September			
Plant	**1**	**2**	**3**	**4**	**1**	**2**	**3**	**4**	**1**	**2**	**3**	**4**

Harvest Times

Month/Week	October				November				December			
Plant	**1**	**2**	**3**	**4**	**1**	**2**	**3**	**4**	**1**	**2**	**3**	**4**

January Harvest

Week 1

Week 2

Week 3

Week 4

February Harvest

Week 1

Week 2

Week 3

Week 4

March Harvest

Week 1

Week 2

Week 3

Week 4

April Harvest

Week 1

Week 2

Week 3

Week 4

May Harvest

Week 1

Week 2

Week 3

Week 4

June Harvest

Week 1

Week 2

Week 3

Week 4

July Harvest

Week 1

Week 2

Week 3

Week 4

August Harvest

Week 1

Week 2

Week 3

Week 4

September Harvest

Week 1

Week 2

Week 3

Week 4

October Harvest

Week 1

Week 2

Week 3

Week 4

November Harvest

Week 1

Week 2

Week 3

Week 4

December Harvest

Week 1

Week 2

Week 3

Week 4

Notes

Section 1:

Vegetable Files

Common Name	
Latin Name	
Cultivar	
Type	
Harvest Time	
Propagation	
Source	
Light Preference	
Soil Type	
Water Requirement	
Companion Plants	
Soil Amendments	
Common Diseases	
Pest Pressure	

Notes:

Vegetable Files

Common Name	
Latin Name	
Cultivar	
Type	
Harvest Time	
Propagation	
Source	
Light Preference	
Soil Type	
Water Requirement	
Companion Plants	
Soil Amendments	
Common Diseases	
Pest Pressure	

Notes:

Vegetable Files

Common Name	
Latin Name	
Cultivar	
Type	
Harvest Time	
Propagation	
Source	
Light Preference	
Soil Type	
Water Requirement	
Companion Plants	
Soil Amendments	
Common Diseases	
Pest Pressure	

Notes:

Vegetable Files

Common Name	
Latin Name	
Cultivar	
Type	
Harvest Time	
Propagation	
Source	
Light Preference	
Soil Type	
Water Requirement	
Companion Plants	
Soil Amendments	
Common Diseases	
Pest Pressure	

Notes:

Vegetable Files

Common Name	
Latin Name	
Cultivar	
Type	
Harvest Time	
Propagation	
Source	
Light Preference	
Soil Type	
Water Requirement	
Companion Plants	
Soil Amendments	
Common Diseases	
Pest Pressure	

Notes:

Vegetable Files

Common Name	
Latin Name	
Cultivar	
Type	
Harvest Time	
Propagation	
Source	
Light Preference	
Soil Type	
Water Requirement	
Companion Plants	
Soil Amendments	
Common Diseases	
Pest Pressure	

Notes:

Vegetable Files

Common Name	
Latin Name	
Cultivar	
Type	
Harvest Time	
Propagation	
Source	
Light Preference	
Soil Type	
Water Requirement	
Companion Plants	
Soil Amendments	
Common Diseases	
Pest Pressure	

Notes:

Vegetable Files

Common Name	
Latin Name	
Cultivar	
Type	
Harvest Time	
Propagation	
Source	
Light Preference	
Soil Type	
Water Requirement	
Companion Plants	
Soil Amendments	
Common Diseases	
Pest Pressure	

Notes:

Vegetable Files

Common Name	
Latin Name	
Cultivar	
Type	
Harvest Time	
Propagation	
Source	
Light Preference	
Soil Type	
Water Requirement	
Companion Plants	
Soil Amendments	
Common Diseases	
Pest Pressure	

Notes:

Vegetable Files

Common Name	
Latin Name	
Cultivar	
Type	
Harvest Time	
Propagation	
Source	
Light Preference	
Soil Type	
Water Requirement	
Companion Plants	
Soil Amendments	
Common Diseases	
Pest Pressure	

Notes:

Vegetable Files

Common Name	
Latin Name	
Cultivar	
Type	
Harvest Time	
Propagation	
Source	
Light Preference	
Soil Type	
Water Requirement	
Companion Plants	
Soil Amendments	
Common Diseases	
Pest Pressure	

Notes:

Vegetable Files

Common Name	
Latin Name	
Cultivar	
Type	
Harvest Time	
Propagation	
Source	
Light Preference	
Soil Type	
Water Requirement	
Companion Plants	
Soil Amendments	
Common Diseases	
Pest Pressure	

Notes:

Notes

Section 2:

Vegetable Files

Common Name	
Latin Name	
Cultivar	
Type	
Harvest Time	
Propagation	
Source	
Light Preference	
Soil Type	
Water Requirement	
Companion Plants	
Soil Amendments	
Common Diseases	
Pest Pressure	

Notes:

Vegetable Files

Common Name	
Latin Name	
Cultivar	
Type	
Harvest Time	
Propagation	
Source	
Light Preference	
Soil Type	
Water Requirement	
Companion Plants	
Soil Amendments	
Common Diseases	
Pest Pressure	

Notes:

Vegetable Files

Common Name	
Latin Name	
Cultivar	
Type	
Harvest Time	
Propagation	
Source	
Light Preference	
Soil Type	
Water Requirement	
Companion Plants	
Soil Amendments	
Common Diseases	
Pest Pressure	

Notes:

Vegetable Files

Common Name	
Latin Name	
Cultivar	
Type	
Harvest Time	
Propagation	
Source	
Light Preference	
Soil Type	
Water Requirement	
Companion Plants	
Soil Amendments	
Common Diseases	
Pest Pressure	

Notes:

Vegetable Files

Common Name	
Latin Name	
Cultivar	
Type	
Harvest Time	
Propagation	
Source	
Light Preference	
Soil Type	
Water Requirement	
Companion Plants	
Soil Amendments	
Common Diseases	
Pest Pressure	

Notes:

Vegetable Files

Common Name	
Latin Name	
Cultivar	
Type	
Harvest Time	
Propagation	
Source	
Light Preference	
Soil Type	
Water Requirement	
Companion Plants	
Soil Amendments	
Common Diseases	
Pest Pressure	

Notes:

Vegetable Files

Common Name	
Latin Name	
Cultivar	
Type	
Harvest Time	
Propagation	
Source	
Light Preference	
Soil Type	
Water Requirement	
Companion Plants	
Soil Amendments	
Common Diseases	
Pest Pressure	

Notes:

Vegetable Files

Common Name	
Latin Name	
Cultivar	
Type	
Harvest Time	
Propagation	
Source	
Light Preference	
Soil Type	
Water Requirement	
Companion Plants	
Soil Amendments	
Common Diseases	
Pest Pressure	

Notes:

Vegetable Files

Common Name	
Latin Name	
Cultivar	
Type	
Harvest Time	
Propagation	
Source	
Light Preference	
Soil Type	
Water Requirement	
Companion Plants	
Soil Amendments	
Common Diseases	
Pest Pressure	

Notes:

Vegetable Files

Common Name	
Latin Name	
Cultivar	
Type	
Harvest Time	
Propagation	
Source	
Light Preference	
Soil Type	
Water Requirement	
Companion Plants	
Soil Amendments	
Common Diseases	
Pest Pressure	

Notes:

Notes

Section 3:

Vegetable Files

Common Name	
Latin Name	
Cultivar	
Type	
Harvest Time	
Propagation	
Source	
Light Preference	
Soil Type	
Water Requirement	
Companion Plants	
Soil Amendments	
Common Diseases	
Pest Pressure	

Notes:

Vegetable Files

Common Name	
Latin Name	
Cultivar	
Type	
Harvest Time	
Propagation	
Source	
Light Preference	
Soil Type	
Water Requirement	
Companion Plants	
Soil Amendments	
Common Diseases	
Pest Pressure	

Notes:

Vegetable Files

Common Name	
Latin Name	
Cultivar	
Type	
Harvest Time	
Propagation	
Source	
Light Preference	
Soil Type	
Water Requirement	
Companion Plants	
Soil Amendments	
Common Diseases	
Pest Pressure	

Notes:

Vegetable Files

Common Name	
Latin Name	
Cultivar	
Type	
Harvest Time	
Propagation	
Source	
Light Preference	
Soil Type	
Water Requirement	
Companion Plants	
Soil Amendments	
Common Diseases	
Pest Pressure	

Notes:

Vegetable Files

Common Name	
Latin Name	
Cultivar	
Type	
Harvest Time	
Propagation	
Source	
Light Preference	
Soil Type	
Water Requirement	
Companion Plants	
Soil Amendments	
Common Diseases	
Pest Pressure	

Notes:

Vegetable Files

Common Name	
Latin Name	
Cultivar	
Type	
Harvest Time	
Propagation	
Source	
Light Preference	
Soil Type	
Water Requirement	
Companion Plants	
Soil Amendments	
Common Diseases	
Pest Pressure	

Notes:

Vegetable Files

Common Name	
Latin Name	
Cultivar	
Type	
Harvest Time	
Propagation	
Source	
Light Preference	
Soil Type	
Water Requirement	
Companion Plants	
Soil Amendments	
Common Diseases	
Pest Pressure	

Notes:

Vegetable Files

Common Name	
Latin Name	
Cultivar	
Type	
Harvest Time	
Propagation	
Source	
Light Preference	
Soil Type	
Water Requirement	
Companion Plants	
Soil Amendments	
Common Diseases	
Pest Pressure	

Notes:

Vegetable Files

Common Name	
Latin Name	
Cultivar	
Type	
Harvest Time	
Propagation	
Source	
Light Preference	
Soil Type	
Water Requirement	
Companion Plants	
Soil Amendments	
Common Diseases	
Pest Pressure	

Notes:

Vegetable Files

Common Name	
Latin Name	
Cultivar	
Type	
Harvest Time	
Propagation	
Source	
Light Preference	
Soil Type	
Water Requirement	
Companion Plants	
Soil Amendments	
Common Diseases	
Pest Pressure	

Notes:

Vegetable Files

Common Name	
Latin Name	
Cultivar	
Type	
Harvest Time	
Propagation	
Source	
Light Preference	
Soil Type	
Water Requirement	
Companion Plants	
Soil Amendments	
Common Diseases	
Pest Pressure	

Notes:

Vegetable Files

Common Name	
Latin Name	
Cultivar	
Type	
Harvest Time	
Propagation	
Source	
Light Preference	
Soil Type	
Water Requirement	
Companion Plants	
Soil Amendments	
Common Diseases	
Pest Pressure	

Notes:

Notes

Section 4:

Vegetable Files

Common Name	
Latin Name	
Cultivar	
Type	
Harvest Time	
Propagation	
Source	
Light Preference	
Soil Type	
Water Requirement	
Companion Plants	
Soil Amendments	
Common Diseases	
Pest Pressure	

Notes:

Vegetable Files

Common Name	
Latin Name	
Cultivar	
Type	
Harvest Time	
Propagation	
Source	
Light Preference	
Soil Type	
Water Requirement	
Companion Plants	
Soil Amendments	
Common Diseases	
Pest Pressure	

Notes:

Vegetable Files

Common Name	
Latin Name	
Cultivar	
Type	
Harvest Time	
Propagation	
Source	
Light Preference	
Soil Type	
Water Requirement	
Companion Plants	
Soil Amendments	
Common Diseases	
Pest Pressure	

Notes:

Vegetable Files

Common Name	
Latin Name	
Cultivar	
Type	
Harvest Time	
Propagation	
Source	
Light Preference	
Soil Type	
Water Requirement	
Companion Plants	
Soil Amendments	
Common Diseases	
Pest Pressure	

Notes:

Vegetable Files

Common Name	
Latin Name	
Cultivar	
Type	
Harvest Time	
Propagation	
Source	
Light Preference	
Soil Type	
Water Requirement	
Companion Plants	
Soil Amendments	
Common Diseases	
Pest Pressure	

Notes:

Vegetable Files

Common Name	
Latin Name	
Cultivar	
Type	
Harvest Time	
Propagation	
Source	
Light Preference	
Soil Type	
Water Requirement	
Companion Plants	
Soil Amendments	
Common Diseases	
Pest Pressure	

Notes:

Vegetable Files

Common Name	
Latin Name	
Cultivar	
Type	
Harvest Time	
Propagation	
Source	
Light Preference	
Soil Type	
Water Requirement	
Companion Plants	
Soil Amendments	
Common Diseases	
Pest Pressure	

Notes:

Vegetable Files

Common Name	
Latin Name	
Cultivar	
Type	
Harvest Time	
Propagation	
Source	
Light Preference	
Soil Type	
Water Requirement	
Companion Plants	
Soil Amendments	
Common Diseases	
Pest Pressure	

Notes:

Vegetable Files

Common Name	
Latin Name	
Cultivar	
Type	
Harvest Time	
Propagation	
Source	
Light Preference	
Soil Type	
Water Requirement	
Companion Plants	
Soil Amendments	
Common Diseases	
Pest Pressure	

Notes:

Vegetable Files

Common Name	
Latin Name	
Cultivar	
Type	
Harvest Time	
Propagation	
Source	
Light Preference	
Soil Type	
Water Requirement	
Companion Plants	
Soil Amendments	
Common Diseases	
Pest Pressure	

Notes:

Vegetable Files

Common Name	
Latin Name	
Cultivar	
Type	
Harvest Time	
Propagation	
Source	
Light Preference	
Soil Type	
Water Requirement	
Companion Plants	
Soil Amendments	
Common Diseases	
Pest Pressure	

Notes:

Vegetable Files

Common Name	
Latin Name	
Cultivar	
Type	
Harvest Time	
Propagation	
Source	
Light Preference	
Soil Type	
Water Requirement	
Companion Plants	
Soil Amendments	
Common Diseases	
Pest Pressure	

Notes:

Notes

Section 5:

Vegetable Files

Common Name	
Latin Name	
Cultivar	
Type	
Harvest Time	
Propagation	
Source	
Light Preference	
Soil Type	
Water Requirement	
Companion Plants	
Soil Amendments	
Common Diseases	
Pest Pressure	

Notes:

Vegetable Files

Common Name	
Latin Name	
Cultivar	
Type	
Harvest Time	
Propagation	
Source	
Light Preference	
Soil Type	
Water Requirement	
Companion Plants	
Soil Amendments	
Common Diseases	
Pest Pressure	

Notes:

Vegetable Files

Common Name	
Latin Name	
Cultivar	
Type	
Harvest Time	
Propagation	
Source	
Light Preference	
Soil Type	
Water Requirement	
Companion Plants	
Soil Amendments	
Common Diseases	
Pest Pressure	

Notes:

Vegetable Files

Common Name	
Latin Name	
Cultivar	
Type	
Harvest Time	
Propagation	
Source	
Light Preference	
Soil Type	
Water Requirement	
Companion Plants	
Soil Amendments	
Common Diseases	
Pest Pressure	

Notes:

Vegetable Files

Common Name	
Latin Name	
Cultivar	
Type	
Harvest Time	
Propagation	
Source	
Light Preference	
Soil Type	
Water Requirement	
Companion Plants	
Soil Amendments	
Common Diseases	
Pest Pressure	

Notes:

Vegetable Files

Common Name	
Latin Name	
Cultivar	
Type	
Harvest Time	
Propagation	
Source	
Light Preference	
Soil Type	
Water Requirement	
Companion Plants	
Soil Amendments	
Common Diseases	
Pest Pressure	

Notes:

Vegetable Files

Common Name	
Latin Name	
Cultivar	
Type	
Harvest Time	
Propagation	
Source	
Light Preference	
Soil Type	
Water Requirement	
Companion Plants	
Soil Amendments	
Common Diseases	
Pest Pressure	

Notes:

Vegetable Files

Common Name	
Latin Name	
Cultivar	
Type	
Harvest Time	
Propagation	
Source	
Light Preference	
Soil Type	
Water Requirement	
Companion Plants	
Soil Amendments	
Common Diseases	
Pest Pressure	

Notes:

Vegetable Files

Common Name	
Latin Name	
Cultivar	
Type	
Harvest Time	
Propagation	
Source	
Light Preference	
Soil Type	
Water Requirement	
Companion Plants	
Soil Amendments	
Common Diseases	
Pest Pressure	

Notes:

Vegetable Files

Common Name	
Latin Name	
Cultivar	
Type	
Harvest Time	
Propagation	
Source	
Light Preference	
Soil Type	
Water Requirement	
Companion Plants	
Soil Amendments	
Common Diseases	
Pest Pressure	

Notes:

Vegetable Files

Common Name	
Latin Name	
Cultivar	
Type	
Harvest Time	
Propagation	
Source	
Light Preference	
Soil Type	
Water Requirement	
Companion Plants	
Soil Amendments	
Common Diseases	
Pest Pressure	

Notes:

Vegetable Files

Common Name	
Latin Name	
Cultivar	
Type	
Harvest Time	
Propagation	
Source	
Light Preference	
Soil Type	
Water Requirement	
Companion Plants	
Soil Amendments	
Common Diseases	
Pest Pressure	

Notes:

Notes

Section 6:

Vegetable Files

Common Name	
Latin Name	
Cultivar	
Type	
Harvest Time	
Propagation	
Source	
Light Preference	
Soil Type	
Water Requirement	
Companion Plants	
Soil Amendments	
Common Diseases	
Pest Pressure	

Notes:

Vegetable Files

Common Name	
Latin Name	
Cultivar	
Type	
Harvest Time	
Propagation	
Source	
Light Preference	
Soil Type	
Water Requirement	
Companion Plants	
Soil Amendments	
Common Diseases	
Pest Pressure	

Notes:

Vegetable Files

Common Name	
Latin Name	
Cultivar	
Type	
Harvest Time	
Propagation	
Source	
Light Preference	
Soil Type	
Water Requirement	
Companion Plants	
Soil Amendments	
Common Diseases	
Pest Pressure	

Notes:

Vegetable Files

Common Name	
Latin Name	
Cultivar	
Type	
Harvest Time	
Propagation	
Source	
Light Preference	
Soil Type	
Water Requirement	
Companion Plants	
Soil Amendments	
Common Diseases	
Pest Pressure	

Notes:

Vegetable Files

Common Name	
Latin Name	
Cultivar	
Type	
Harvest Time	
Propagation	
Source	
Light Preference	
Soil Type	
Water Requirement	
Companion Plants	
Soil Amendments	
Common Diseases	
Pest Pressure	

Notes:

Vegetable Files

Common Name	
Latin Name	
Cultivar	
Type	
Harvest Time	
Propagation	
Source	
Light Preference	
Soil Type	
Water Requirement	
Companion Plants	
Soil Amendments	
Common Diseases	
Pest Pressure	

Notes:

Vegetable Files

Common Name	
Latin Name	
Cultivar	
Type	
Harvest Time	
Propagation	
Source	
Light Preference	
Soil Type	
Water Requirement	
Companion Plants	
Soil Amendments	
Common Diseases	
Pest Pressure	

Notes:

Vegetable Files

Common Name	
Latin Name	
Cultivar	
Type	
Harvest Time	
Propagation	
Source	
Light Preference	
Soil Type	
Water Requirement	
Companion Plants	
Soil Amendments	
Common Diseases	
Pest Pressure	

Notes:

Vegetable Files

Common Name	
Latin Name	
Cultivar	
Type	
Harvest Time	
Propagation	
Source	
Light Preference	
Soil Type	
Water Requirement	
Companion Plants	
Soil Amendments	
Common Diseases	
Pest Pressure	

Notes:

Vegetable Files

Common Name	
Latin Name	
Cultivar	
Type	
Harvest Time	
Propagation	
Source	
Light Preference	
Soil Type	
Water Requirement	
Companion Plants	
Soil Amendments	
Common Diseases	
Pest Pressure	

Notes:

Vegetable Files

Common Name	
Latin Name	
Cultivar	
Type	
Harvest Time	
Propagation	
Source	
Light Preference	
Soil Type	
Water Requirement	
Companion Plants	
Soil Amendments	
Common Diseases	
Pest Pressure	

Notes:

Vegetable Files

Common Name	
Latin Name	
Cultivar	
Type	
Harvest Time	
Propagation	
Source	
Light Preference	
Soil Type	
Water Requirement	
Companion Plants	
Soil Amendments	
Common Diseases	
Pest Pressure	

Notes:

Notes

Section 7:

Vegetable Files

Common Name	
Latin Name	
Cultivar	
Type	
Harvest Time	
Propagation	
Source	
Light Preference	
Soil Type	
Water Requirement	
Companion Plants	
Soil Amendments	
Common Diseases	
Pest Pressure	

Notes:

Vegetable Files

Common Name	
Latin Name	
Cultivar	
Type	
Harvest Time	
Propagation	
Source	
Light Preference	
Soil Type	
Water Requirement	
Companion Plants	
Soil Amendments	
Common Diseases	
Pest Pressure	

Notes:

Vegetable Files

Common Name	
Latin Name	
Cultivar	
Type	
Harvest Time	
Propagation	
Source	
Light Preference	
Soil Type	
Water Requirement	
Companion Plants	
Soil Amendments	
Common Diseases	
Pest Pressure	

Notes:

Vegetable Files

Common Name	
Latin Name	
Cultivar	
Type	
Harvest Time	
Propagation	
Source	
Light Preference	
Soil Type	
Water Requirement	
Companion Plants	
Soil Amendments	
Common Diseases	
Pest Pressure	

Notes:

Vegetable Files

Common Name	
Latin Name	
Cultivar	
Type	
Harvest Time	
Propagation	
Source	
Light Preference	
Soil Type	
Water Requirement	
Companion Plants	
Soil Amendments	
Common Diseases	
Pest Pressure	

Notes:

Vegetable Files

Common Name	
Latin Name	
Cultivar	
Type	
Harvest Time	
Propagation	
Source	
Light Preference	
Soil Type	
Water Requirement	
Companion Plants	
Soil Amendments	
Common Diseases	
Pest Pressure	

Notes:

Vegetable Files

Common Name	
Latin Name	
Cultivar	
Type	
Harvest Time	
Propagation	
Source	
Light Preference	
Soil Type	
Water Requirement	
Companion Plants	
Soil Amendments	
Common Diseases	
Pest Pressure	

Notes:

Vegetable Files

Common Name	
Latin Name	
Cultivar	
Type	
Harvest Time	
Propagation	
Source	
Light Preference	
Soil Type	
Water Requirement	
Companion Plants	
Soil Amendments	
Common Diseases	
Pest Pressure	

Notes:

Vegetable Files

Common Name	
Latin Name	
Cultivar	
Type	
Harvest Time	
Propagation	
Source	
Light Preference	
Soil Type	
Water Requirement	
Companion Plants	
Soil Amendments	
Common Diseases	
Pest Pressure	

Notes:

Vegetable Files

Common Name	
Latin Name	
Cultivar	
Type	
Harvest Time	
Propagation	
Source	
Light Preference	
Soil Type	
Water Requirement	
Companion Plants	
Soil Amendments	
Common Diseases	
Pest Pressure	

Notes:

Vegetable Files

Common Name	
Latin Name	
Cultivar	
Type	
Harvest Time	
Propagation	
Source	
Light Preference	
Soil Type	
Water Requirement	
Companion Plants	
Soil Amendments	
Common Diseases	
Pest Pressure	

Notes:

Vegetable Files

Common Name	
Latin Name	
Cultivar	
Type	
Harvest Time	
Propagation	
Source	
Light Preference	
Soil Type	
Water Requirement	
Companion Plants	
Soil Amendments	
Common Diseases	
Pest Pressure	

Notes:

Notes

Section 8:

Vegetable Files

Common Name	
Latin Name	
Cultivar	
Type	
Harvest Time	
Propagation	
Source	
Light Preference	
Soil Type	
Water Requirement	
Companion Plants	
Soil Amendments	
Common Diseases	
Pest Pressure	

Notes:

Vegetable Files

Common Name	
Latin Name	
Cultivar	
Type	
Harvest Time	
Propagation	
Source	
Light Preference	
Soil Type	
Water Requirement	
Companion Plants	
Soil Amendments	
Common Diseases	
Pest Pressure	

Notes:

Vegetable Files

Common Name	
Latin Name	
Cultivar	
Type	
Harvest Time	
Propagation	
Source	
Light Preference	
Soil Type	
Water Requirement	
Companion Plants	
Soil Amendments	
Common Diseases	
Pest Pressure	

Notes:

Vegetable Files

Common Name	
Latin Name	
Cultivar	
Type	
Harvest Time	
Propagation	
Source	
Light Preference	
Soil Type	
Water Requirement	
Companion Plants	
Soil Amendments	
Common Diseases	
Pest Pressure	

Notes:

Vegetable Files

Common Name	
Latin Name	
Cultivar	
Type	
Harvest Time	
Propagation	
Source	
Light Preference	
Soil Type	
Water Requirement	
Companion Plants	
Soil Amendments	
Common Diseases	
Pest Pressure	

Notes:

Vegetable Files

Common Name	
Latin Name	
Cultivar	
Type	
Harvest Time	
Propagation	
Source	
Light Preference	
Soil Type	
Water Requirement	
Companion Plants	
Soil Amendments	
Common Diseases	
Pest Pressure	

Notes:

Vegetable Files

Common Name	
Latin Name	
Cultivar	
Type	
Harvest Time	
Propagation	
Source	
Light Preference	
Soil Type	
Water Requirement	
Companion Plants	
Soil Amendments	
Common Diseases	
Pest Pressure	

Notes:

Vegetable Files

Common Name	
Latin Name	
Cultivar	
Type	
Harvest Time	
Propagation	
Source	
Light Preference	
Soil Type	
Water Requirement	
Companion Plants	
Soil Amendments	
Common Diseases	
Pest Pressure	

Notes:

Vegetable Files

Common Name	
Latin Name	
Cultivar	
Type	
Harvest Time	
Propagation	
Source	
Light Preference	
Soil Type	
Water Requirement	
Companion Plants	
Soil Amendments	
Common Diseases	
Pest Pressure	

Notes:

Vegetable Files

Common Name	
Latin Name	
Cultivar	
Type	
Harvest Time	
Propagation	
Source	
Light Preference	
Soil Type	
Water Requirement	
Companion Plants	
Soil Amendments	
Common Diseases	
Pest Pressure	

Notes:

Vegetable Files

Common Name	
Latin Name	
Cultivar	
Type	
Harvest Time	
Propagation	
Source	
Light Preference	
Soil Type	
Water Requirement	
Companion Plants	
Soil Amendments	
Common Diseases	
Pest Pressure	

Notes:

Vegetable Files

Common Name	
Latin Name	
Cultivar	
Type	
Harvest Time	
Propagation	
Source	
Light Preference	
Soil Type	
Water Requirement	
Companion Plants	
Soil Amendments	
Common Diseases	
Pest Pressure	

Notes:

Notes

Section 9:

Vegetable Files

Common Name	
Latin Name	
Cultivar	
Type	
Harvest Time	
Propagation	
Source	
Light Preference	
Soil Type	
Water Requirement	
Companion Plants	
Soil Amendments	
Common Diseases	
Pest Pressure	

Notes:

Vegetable Files

Common Name	
Latin Name	
Cultivar	
Type	
Harvest Time	
Propagation	
Source	
Light Preference	
Soil Type	
Water Requirement	
Companion Plants	
Soil Amendments	
Common Diseases	
Pest Pressure	

Notes:

Vegetable Files

Common Name	
Latin Name	
Cultivar	
Type	
Harvest Time	
Propagation	
Source	
Light Preference	
Soil Type	
Water Requirement	
Companion Plants	
Soil Amendments	
Common Diseases	
Pest Pressure	

Notes:

Vegetable Files

Common Name	
Latin Name	
Cultivar	
Type	
Harvest Time	
Propagation	
Source	
Light Preference	
Soil Type	
Water Requirement	
Companion Plants	
Soil Amendments	
Common Diseases	
Pest Pressure	

Notes:

Vegetable Files

Common Name	
Latin Name	
Cultivar	
Type	
Harvest Time	
Propagation	
Source	
Light Preference	
Soil Type	
Water Requirement	
Companion Plants	
Soil Amendments	
Common Diseases	
Pest Pressure	

Notes:

Vegetable Files

Common Name	
Latin Name	
Cultivar	
Type	
Harvest Time	
Propagation	
Source	
Light Preference	
Soil Type	
Water Requirement	
Companion Plants	
Soil Amendments	
Common Diseases	
Pest Pressure	

Notes:

Vegetable Files

Common Name	
Latin Name	
Cultivar	
Type	
Harvest Time	
Propagation	
Source	
Light Preference	
Soil Type	
Water Requirement	
Companion Plants	
Soil Amendments	
Common Diseases	
Pest Pressure	

Notes:

Vegetable Files

Common Name	
Latin Name	
Cultivar	
Type	
Harvest Time	
Propagation	
Source	
Light Preference	
Soil Type	
Water Requirement	
Companion Plants	
Soil Amendments	
Common Diseases	
Pest Pressure	

Notes:

Vegetable Files

Common Name	
Latin Name	
Cultivar	
Type	
Harvest Time	
Propagation	
Source	
Light Preference	
Soil Type	
Water Requirement	
Companion Plants	
Soil Amendments	
Common Diseases	
Pest Pressure	

Notes:

Vegetable Files

Common Name	
Latin Name	
Cultivar	
Type	
Harvest Time	
Propagation	
Source	
Light Preference	
Soil Type	
Water Requirement	
Companion Plants	
Soil Amendments	
Common Diseases	
Pest Pressure	

Notes:

Vegetable Files

Common Name	
Latin Name	
Cultivar	
Type	
Harvest Time	
Propagation	
Source	
Light Preference	
Soil Type	
Water Requirement	
Companion Plants	
Soil Amendments	
Common Diseases	
Pest Pressure	

Notes:

Vegetable Files

Common Name	
Latin Name	
Cultivar	
Type	
Harvest Time	
Propagation	
Source	
Light Preference	
Soil Type	
Water Requirement	
Companion Plants	
Soil Amendments	
Common Diseases	
Pest Pressure	

Notes:

Notes

Section 10:

Vegetable Files

Common Name	
Latin Name	
Cultivar	
Type	
Harvest Time	
Propagation	
Source	
Light Preference	
Soil Type	
Water Requirement	
Companion Plants	
Soil Amendments	
Common Diseases	
Pest Pressure	

Notes:

Vegetable Files

Common Name	
Latin Name	
Cultivar	
Type	
Harvest Time	
Propagation	
Source	
Light Preference	
Soil Type	
Water Requirement	
Companion Plants	
Soil Amendments	
Common Diseases	
Pest Pressure	

Notes:

Vegetable Files

Common Name	
Latin Name	
Cultivar	
Type	
Harvest Time	
Propagation	
Source	
Light Preference	
Soil Type	
Water Requirement	
Companion Plants	
Soil Amendments	
Common Diseases	
Pest Pressure	

Notes:

Vegetable Files

Common Name	
Latin Name	
Cultivar	
Type	
Harvest Time	
Propagation	
Source	
Light Preference	
Soil Type	
Water Requirement	
Companion Plants	
Soil Amendments	
Common Diseases	
Pest Pressure	

Notes:

Vegetable Files

Common Name	
Latin Name	
Cultivar	
Type	
Harvest Time	
Propagation	
Source	
Light Preference	
Soil Type	
Water Requirement	
Companion Plants	
Soil Amendments	
Common Diseases	
Pest Pressure	

Notes:

Vegetable Files

Common Name	
Latin Name	
Cultivar	
Type	
Harvest Time	
Propagation	
Source	
Light Preference	
Soil Type	
Water Requirement	
Companion Plants	
Soil Amendments	
Common Diseases	
Pest Pressure	

Notes:

Vegetable Files

Common Name	
Latin Name	
Cultivar	
Type	
Harvest Time	
Propagation	
Source	
Light Preference	
Soil Type	
Water Requirement	
Companion Plants	
Soil Amendments	
Common Diseases	
Pest Pressure	

Notes:

Vegetable Files

Common Name	
Latin Name	
Cultivar	
Type	
Harvest Time	
Propagation	
Source	
Light Preference	
Soil Type	
Water Requirement	
Companion Plants	
Soil Amendments	
Common Diseases	
Pest Pressure	

Notes:

Vegetable Files

Common Name	
Latin Name	
Cultivar	
Type	
Harvest Time	
Propagation	
Source	
Light Preference	
Soil Type	
Water Requirement	
Companion Plants	
Soil Amendments	
Common Diseases	
Pest Pressure	

Notes:

Vegetable Files

Common Name	
Latin Name	
Cultivar	
Type	
Harvest Time	
Propagation	
Source	
Light Preference	
Soil Type	
Water Requirement	
Companion Plants	
Soil Amendments	
Common Diseases	
Pest Pressure	

Notes:

Vegetable Files

Common Name	
Latin Name	
Cultivar	
Type	
Harvest Time	
Propagation	
Source	
Light Preference	
Soil Type	
Water Requirement	
Companion Plants	
Soil Amendments	
Common Diseases	
Pest Pressure	

Notes:

Common Name	
Latin Name	
Cultivar	
Type	
Harvest Time	
Propagation	
Source	
Light Preference	
Soil Type	
Water Requirement	
Companion Plants	
Soil Amendments	
Common Diseases	
Pest Pressure	

Notes:

Notes

Section 11:

Vegetable Files

Common Name	
Latin Name	
Cultivar	
Type	
Harvest Time	
Propagation	
Source	
Light Preference	
Soil Type	
Water Requirement	
Companion Plants	
Soil Amendments	
Common Diseases	
Pest Pressure	

Notes:

Vegetable Files

Common Name	
Latin Name	
Cultivar	
Type	
Harvest Time	
Propagation	
Source	
Light Preference	
Soil Type	
Water Requirement	
Companion Plants	
Soil Amendments	
Common Diseases	
Pest Pressure	

Notes:

Vegetable Files

Common Name	
Latin Name	
Cultivar	
Type	
Harvest Time	
Propagation	
Source	
Light Preference	
Soil Type	
Water Requirement	
Companion Plants	
Soil Amendments	
Common Diseases	
Pest Pressure	

Notes:

Vegetable Files

Common Name	
Latin Name	
Cultivar	
Type	
Harvest Time	
Propagation	
Source	
Light Preference	
Soil Type	
Water Requirement	
Companion Plants	
Soil Amendments	
Common Diseases	
Pest Pressure	

Notes:

Vegetable Files

Common Name	
Latin Name	
Cultivar	
Type	
Harvest Time	
Propagation	
Source	
Light Preference	
Soil Type	
Water Requirement	
Companion Plants	
Soil Amendments	
Common Diseases	
Pest Pressure	

Notes:

Vegetable Files

Common Name	
Latin Name	
Cultivar	
Type	
Harvest Time	
Propagation	
Source	
Light Preference	
Soil Type	
Water Requirement	
Companion Plants	
Soil Amendments	
Common Diseases	
Pest Pressure	

Notes:

Vegetable Files

Common Name	
Latin Name	
Cultivar	
Type	
Harvest Time	
Propagation	
Source	
Light Preference	
Soil Type	
Water Requirement	
Companion Plants	
Soil Amendments	
Common Diseases	
Pest Pressure	

Notes:

Vegetable Files

Common Name	
Latin Name	
Cultivar	
Type	
Harvest Time	
Propagation	
Source	
Light Preference	
Soil Type	
Water Requirement	
Companion Plants	
Soil Amendments	
Common Diseases	
Pest Pressure	

Notes:

Vegetable Files

Common Name	
Latin Name	
Cultivar	
Type	
Harvest Time	
Propagation	
Source	
Light Preference	
Soil Type	
Water Requirement	
Companion Plants	
Soil Amendments	
Common Diseases	
Pest Pressure	

Notes:

Vegetable Files

Common Name	
Latin Name	
Cultivar	
Type	
Harvest Time	
Propagation	
Source	
Light Preference	
Soil Type	
Water Requirement	
Companion Plants	
Soil Amendments	
Common Diseases	
Pest Pressure	

Notes:

Vegetable Files

Common Name	
Latin Name	
Cultivar	
Type	
Harvest Time	
Propagation	
Source	
Light Preference	
Soil Type	
Water Requirement	
Companion Plants	
Soil Amendments	
Common Diseases	
Pest Pressure	

Notes:

Vegetable Files

Common Name	
Latin Name	
Cultivar	
Type	
Harvest Time	
Propagation	
Source	
Light Preference	
Soil Type	
Water Requirement	
Companion Plants	
Soil Amendments	
Common Diseases	
Pest Pressure	

Notes:

Notes

Section 12:

Vegetable Files

Common Name	
Latin Name	
Cultivar	
Type	
Harvest Time	
Propagation	
Source	
Light Preference	
Soil Type	
Water Requirement	
Companion Plants	
Soil Amendments	
Common Diseases	
Pest Pressure	

Notes:

Vegetable Files

Common Name	
Latin Name	
Cultivar	
Type	
Harvest Time	
Propagation	
Source	
Light Preference	
Soil Type	
Water Requirement	
Companion Plants	
Soil Amendments	
Common Diseases	
Pest Pressure	

Notes:

Vegetable Files

Common Name	
Latin Name	
Cultivar	
Type	
Harvest Time	
Propagation	
Source	
Light Preference	
Soil Type	
Water Requirement	
Companion Plants	
Soil Amendments	
Common Diseases	
Pest Pressure	

Notes:

Vegetable Files

Common Name	
Latin Name	
Cultivar	
Type	
Harvest Time	
Propagation	
Source	
Light Preference	
Soil Type	
Water Requirement	
Companion Plants	
Soil Amendments	
Common Diseases	
Pest Pressure	

Notes:

Vegetable Files

Common Name	
Latin Name	
Cultivar	
Type	
Harvest Time	
Propagation	
Source	
Light Preference	
Soil Type	
Water Requirement	
Companion Plants	
Soil Amendments	
Common Diseases	
Pest Pressure	

Notes:

Vegetable Files

Common Name	
Latin Name	
Cultivar	
Type	
Harvest Time	
Propagation	
Source	
Light Preference	
Soil Type	
Water Requirement	
Companion Plants	
Soil Amendments	
Common Diseases	
Pest Pressure	

Notes:

Vegetable Files

Common Name	
Latin Name	
Cultivar	
Type	
Harvest Time	
Propagation	
Source	
Light Preference	
Soil Type	
Water Requirement	
Companion Plants	
Soil Amendments	
Common Diseases	
Pest Pressure	

Notes:

Vegetable Files

Common Name	
Latin Name	
Cultivar	
Type	
Harvest Time	
Propagation	
Source	
Light Preference	
Soil Type	
Water Requirement	
Companion Plants	
Soil Amendments	
Common Diseases	
Pest Pressure	

Notes:

Vegetable Files

Common Name	
Latin Name	
Cultivar	
Type	
Harvest Time	
Propagation	
Source	
Light Preference	
Soil Type	
Water Requirement	
Companion Plants	
Soil Amendments	
Common Diseases	
Pest Pressure	

Notes:

Vegetable Files

Common Name	
Latin Name	
Cultivar	
Type	
Harvest Time	
Propagation	
Source	
Light Preference	
Soil Type	
Water Requirement	
Companion Plants	
Soil Amendments	
Common Diseases	
Pest Pressure	

Notes:

Vegetable Files

Common Name	
Latin Name	
Cultivar	
Type	
Harvest Time	
Propagation	
Source	
Light Preference	
Soil Type	
Water Requirement	
Companion Plants	
Soil Amendments	
Common Diseases	
Pest Pressure	

Notes:

Vegetable Files

Common Name	
Latin Name	
Cultivar	
Type	
Harvest Time	
Propagation	
Source	
Light Preference	
Soil Type	
Water Requirement	
Companion Plants	
Soil Amendments	
Common Diseases	
Pest Pressure	

Notes:

Notes

Section 13:

Vegetable Files

Common Name	
Latin Name	
Cultivar	
Type	
Harvest Time	
Propagation	
Source	
Light Preference	
Soil Type	
Water Requirement	
Companion Plants	
Soil Amendments	
Common Diseases	
Pest Pressure	

Notes:

Vegetable Files

Common Name	
Latin Name	
Cultivar	
Type	
Harvest Time	
Propagation	
Source	
Light Preference	
Soil Type	
Water Requirement	
Companion Plants	
Soil Amendments	
Common Diseases	
Pest Pressure	

Notes:

Vegetable Files

Common Name	
Latin Name	
Cultivar	
Type	
Harvest Time	
Propagation	
Source	
Light Preference	
Soil Type	
Water Requirement	
Companion Plants	
Soil Amendments	
Common Diseases	
Pest Pressure	

Notes:

Vegetable Files

Common Name	
Latin Name	
Cultivar	
Type	
Harvest Time	
Propagation	
Source	
Light Preference	
Soil Type	
Water Requirement	
Companion Plants	
Soil Amendments	
Common Diseases	
Pest Pressure	

Notes:

Vegetable Files

Common Name	
Latin Name	
Cultivar	
Type	
Harvest Time	
Propagation	
Source	
Light Preference	
Soil Type	
Water Requirement	
Companion Plants	
Soil Amendments	
Common Diseases	
Pest Pressure	

Notes:

Vegetable Files

Common Name	
Latin Name	
Cultivar	
Type	
Harvest Time	
Propagation	
Source	
Light Preference	
Soil Type	
Water Requirement	
Companion Plants	
Soil Amendments	
Common Diseases	
Pest Pressure	

Notes:

Vegetable Files

Common Name	
Latin Name	
Cultivar	
Type	
Harvest Time	
Propagation	
Source	
Light Preference	
Soil Type	
Water Requirement	
Companion Plants	
Soil Amendments	
Common Diseases	
Pest Pressure	

Notes:

Vegetable Files

Common Name	
Latin Name	
Cultivar	
Type	
Harvest Time	
Propagation	
Source	
Light Preference	
Soil Type	
Water Requirement	
Companion Plants	
Soil Amendments	
Common Diseases	
Pest Pressure	

Notes:

Vegetable Files

Common Name	
Latin Name	
Cultivar	
Type	
Harvest Time	
Propagation	
Source	
Light Preference	
Soil Type	
Water Requirement	
Companion Plants	
Soil Amendments	
Common Diseases	
Pest Pressure	

Notes:

Vegetable Files

Common Name	
Latin Name	
Cultivar	
Type	
Harvest Time	
Propagation	
Source	
Light Preference	
Soil Type	
Water Requirement	
Companion Plants	
Soil Amendments	
Common Diseases	
Pest Pressure	

Notes:

Vegetable Files

Common Name	
Latin Name	
Cultivar	
Type	
Harvest Time	
Propagation	
Source	
Light Preference	
Soil Type	
Water Requirement	
Companion Plants	
Soil Amendments	
Common Diseases	
Pest Pressure	

Notes:

Vegetable Files

Common Name	
Latin Name	
Cultivar	
Type	
Harvest Time	
Propagation	
Source	
Light Preference	
Soil Type	
Water Requirement	
Companion Plants	
Soil Amendments	
Common Diseases	
Pest Pressure	

Notes:

Notes

Section 14:

Vegetable Files

Common Name	
Latin Name	
Cultivar	
Type	
Harvest Time	
Propagation	
Source	
Light Preference	
Soil Type	
Water Requirement	
Companion Plants	
Soil Amendments	
Common Diseases	
Pest Pressure	

Notes:

Vegetable Files

Common Name	
Latin Name	
Cultivar	
Type	
Harvest Time	
Propagation	
Source	
Light Preference	
Soil Type	
Water Requirement	
Companion Plants	
Soil Amendments	
Common Diseases	
Pest Pressure	

Notes:

Vegetable Files

Common Name	
Latin Name	
Cultivar	
Type	
Harvest Time	
Propagation	
Source	
Light Preference	
Soil Type	
Water Requirement	
Companion Plants	
Soil Amendments	
Common Diseases	
Pest Pressure	

Notes:

Vegetable Files

Common Name	
Latin Name	
Cultivar	
Type	
Harvest Time	
Propagation	
Source	
Light Preference	
Soil Type	
Water Requirement	
Companion Plants	
Soil Amendments	
Common Diseases	
Pest Pressure	

Notes:

Vegetable Files

Common Name	
Latin Name	
Cultivar	
Type	
Harvest Time	
Propagation	
Source	
Light Preference	
Soil Type	
Water Requirement	
Companion Plants	
Soil Amendments	
Common Diseases	
Pest Pressure	

Notes:

Vegetable Files

Common Name	
Latin Name	
Cultivar	
Type	
Harvest Time	
Propagation	
Source	
Light Preference	
Soil Type	
Water Requirement	
Companion Plants	
Soil Amendments	
Common Diseases	
Pest Pressure	

Notes:

Notes

Notes

Melanie Holsti is an avid tea lover with a deep appreciation for the simple joys of life. Whether she's sipping a cup of Scottish Breakfast tea, capturing the perfect shot with her camera, or getting her hands dirty in the garden, she embraces each moment with enthusiasm.

As an author and small business owner, she pours her focus and creativity into everything she does.

Her background is as varied as her interests—she's worked in hospitality, retail, livestock farming, writing, and, of course, gardening. Over the years, she has experienced the triumphs and challenges that come with growing food and homesteading in diverse climates, from the harsh prairie of South Dakota to the woods and mountains of Missouri and Tennessee.

In 2018, Melanie and her family took on an exciting project—restoring a charming 1917 Craftsman bungalow. With a lot of hard work and a shared vision, they transformed it into the Ironstone Cottage Tea Room, where guests could gather and enjoy a hot cup of tea or lunch with friends in a cozy, welcoming atmosphere.

These days, Melanie, her husband, and their daughters reside on a small farm in Tennessee, living in a beautiful Federal-style home they built together. Their homestead is shared with an assortment of animals, including a peculiar flock of barn cats, free-ranging chickens, and noisy ducks.

When she's not tending to the garden or caring for her animals, Melanie finds joy in writing books, sampling new and exotic teas, and cheering on her favorite professional hockey team. No matter the season, she's always pursuing her interests and creating a life filled with warmth and, of course, plenty of tea.

www.ironstone_tennessee.com

IronstoneTennessee

IronstoneTennessee

amazon.com/author/ironstonetennessee

ironstonetennessee

secret resources web page
Password: IronstoneBooks

Your feedback and reviews are priceless. If you have suggestions for improving my gardening workbooks, I'd love to hear them. If you have a moment, please consider leaving a review on Amazon.

Other Titles by Melanie Holsti

The Ironstone Garden Planner (color)

The Ironstone Garden Journal(color)

The Vegetable Files

The Houseplant Files

The Herb Files

The Orchard Files

The Fruit, Flower, Herb & Vegetable Garden Planner

The Fruit, Flower, Herb & Vegetable Garden Journal

Journaling Your Weather

The Ironstone Forager's Notebook

Journaling Your Health (various designs)

Ironstone Prayer Journals (various designs)

Church & Assembly Prayer Team Journals (various designs)

Tea & Tisane Blending Workbook: A Tea Mixologist's Journal

(with Jennifer C. Petersen)

Made in the USA
Monee, IL
05 November 2026

67911529R10122